Motorcycle Classics

750 Super
Sport

Grant Leonard

Contents

Published by Magna Books
Magna Road
Wigston
Leicester LE18 4ZH

Produced by Bison Books Ltd
Kimbolton House
117A Fulham Road
London SW3 6RL

ISBN 1-85422-526-X

Reprinted 1993
Reprinted 1994

Printed in China

DUCATI
DESMO

DESMO
900
SUPER SPORT
UOH
900S

Page 1: *A detail of the classic Harley-Davidson XR1000.*

Gatefold: *The charismatic Ducati 750SS (left) and 900SS.*

Page 9: *A Harley-Davidson Tour Glide Classic.*

Acknowledgments

The publisher would like to thank Adrian Hodgkins the designer, Ron Watson for preparing the index and the agencies and individuals listed below for supplying the photographs:

Roland Brown pp 69

Tony Butler pp 36-7

David Goldman pp 35 (bottom), 44, 45, 55 (bottom), 58 (bottom), 60, 64, 66, 67, 68 (top), 73, 76-7

Patrick Gosling pp 49 (both), 50-1, 65 (both)

Tony La Gruth pp 15

Roy Kidney pp 18-19 (all three)

Grant Leonard pp 1, 9, 11 (both), 14, 21 (top), 29, 52, 54, 55 (top), 70

Julian Mackie pp 62, 63

Phil Masters Photography pp gatefold, 48, 57 (top), 79 (top)

Andrew Morland pp 20 (left), 21 (bottom), 24, 25, 27, 28 (bottom), 39, 46, 71, 75 (top), 79 (bottom)

Don Morley pp 10, 12, 13, 16 (both), 17, 22, 23, 26, 28 (top), 31, 34, 38, 43, 47 (top), 53, 68 (bottom), 72, 74, 75 (bottom), 78 (top left and top right)

Superbike pp 30 (both), 35 (top), 40, 41, 42 (both), 58-9, 61 (both)

Oli Tennent pp 47 (bottom)

Harley-Davidson's 'Silent Gray Fellow'

Harley-Davidson was established in 1903 and is today the oldest motorcycle company in the world. That gives their first popular model, the 35-cubic-inch 9.35, a special place in motorcycling history – the grandfather of motorcycles. It earned the monicker 'Silent Gray Fellow' because in the first place its exhaust note was well muffled; secondly, you could have any color you liked as long as it was gray; and the 'Fellow' part? Like the four-legged transport it replaced, the 9.35 proved itself to be a reliable friend to all who chose to own one.

The 9.35 model was belt-driven and single gear, and in 1912 was updated, incorporating an innovation for its time – a clutch. This took a great deal of risk out of the hazardous performance of stop/starting in traffic. It also allowed Bill Harley to use a roller chain for final drive instead of the leather belt, which tended to slip in wet weather.

The 'Silent Gray Fellow' was the basis for the company's development of the V-twin in 1909 which was simply the same engine with a second cylinder grafted on at an angle of 45 degrees – the engine configuration which Harley has maintained ever since that time.

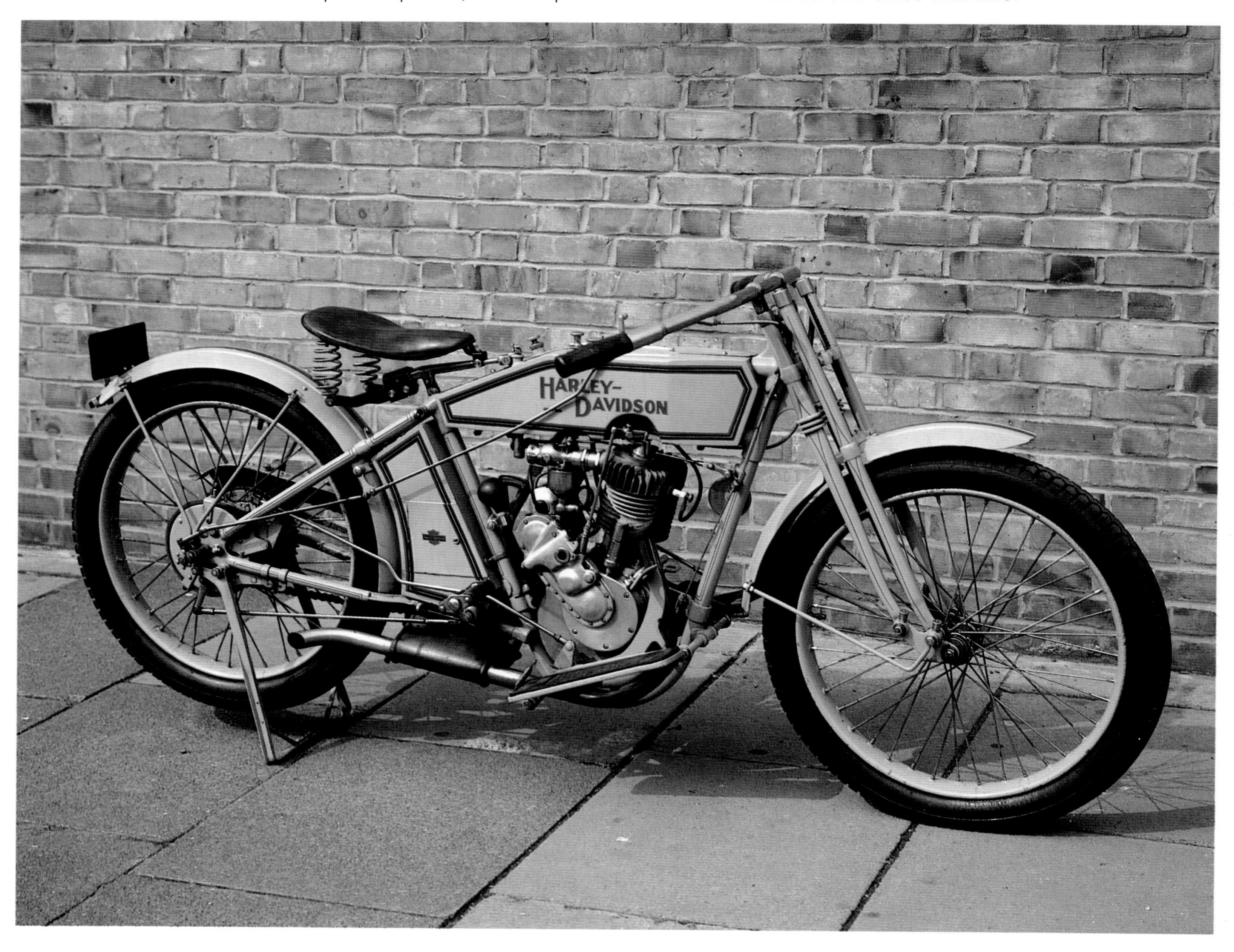

Left and below left: *The Silent Gray Fellow of 1912 featured a leather belt and pulleys for an 'all-or-nothing' direct drive. The detail (below) shows a nine horsepower, 35-cubic-inch single-cylinder motor.*

Indian Scout

The Indian motorcycle was, in its day, the premier American motorcycle, with Harley-Davidson riding on its coat-tails right up until Indian ceased making bikes in 1953. The marque had its origins at the turn of the century when George M Hendee set up a bicycle and motorized bicycle factory in New England. In the seven years leading up to World War I, over 100,000 V-twin Indians had left the so-called 'Wigwam,' the Hendee Manufacturing Company of Springfield, Massachusetts.

The Indian Scout was the brainchild of Charles B Franklin, a longtime Indian engineer and factory racer. The Scout was a middleweight machine of 37 cubic inches (600cc), a V-twin of course, but with a 42-degree cylinder angle and side-by-side valves. It was of unit construction with a three-speed gearbox and 'solid' primary drive (three helical gears, rather than a chain). It produced 12hp, a respectable figure for the day, and top speed was around 55mph.

The Scout was launched in 1920 and began a production life which would span 30 years. In 1926 it was enlarged to 45 cubic inches (750cc) to compete with the Excelsior. Harley would follow, launching their 45 in 1929. The most famous Scout models were the 101, launched in 1929,

Far left: *By 1914, Bill Harley had come up with a clutch mechanism which allowed him to fit a chain drive. It made stopping and starting a much less hazardous exercise.*

Above: *'You can't wear out an Indian Scout.' This model, fitted with a sidecar, shows the old slogan to have been well-founded. It's over 70 years old!*

Far right: *How! A proud owner displays his Indian Four, the final incarnation of the Henderson 58.*

and the Sport Scout (1934). During the 1920s the bike was a phenomenal success and proved so reliable and durable, the factory coined the slogan: 'You can't wear out an Indian Scout.'

The 101 developed over 20hp, weighed 370lb and potentially reached 75mph, but its real forte was its superb handling and balance – such qualities made it the favorite mount of crazy 'wall of death' stunt riders.

Henderson 58

The Henderson 58 is a true classic of American motorcycle design; it is a machine which established an American tradition of 'big is best' in motorcycling. The engine was a gargantuan four-cylinder block, which flew in the face of motorcycle design fashion of the time (and thereafter) which claimed the V-twin as indispensable.

William G Henderson, a Glaswegian emigré living in Rochester, New York, built his prototype in 1910. It was inspired by Belgian Paul Kelecom's design of 1904, popularly known as the FN (Fabrique Nationale D'Armes de Guerre).

Henderson's bike featured many innovations like an overhead inlet and side exhaust valve arrangement. It had magneto ignition and drip oiling system. Displacement was 58 cubic inches, it was single gear with a car-type clutch and was started by means of a crank handle. Henderson leased a factory in the home of the automobile – Detroit, Michigan – and in 1912 the initial production run of 1000 bikes sold out quickly at a bargain price of $325.

The bike remained in production until 1918 when it featured a shorter wheelbase (58 inches), wet-sump lubrication and a three-speed gearbox, all along the lines of current thinking in automobile engineering. In 1918 one Alan Bedell traveled from Los Angeles to New York – 3296 miles – in just seven days and sixteen hours aboard a Henderson four. The previous record of four years' standing, held by an Indian-mounted rider, was beaten by almost four days!

The marque disappeared that same year when Ignatz Schwinn, owner of Excelsior, bought the design rights for a royal sum. Although Henderson, with his brother Tom, went to work for Excelsior with his design, they both resigned within months when Schwinn insisted on redesigning the machine along more economical lines. By no means, though, was it the end for the Henderson motorcycle.

While Excelsior were at the drawing board designing the new bike based on the Henderson (but with an 80-cubic-inch displacement), William Henderson was moving quickly. He set up shop again and designed a new 75-cubic-inch four, which he launched as the Ace in 1920, priced competitively at $375. It wasn't until 1923 that the Excelsior company produced the Henderson De Luxe, billed as the World's Finest Motorcycle.

Tragedy struck the new Ace marque in 1923, when Henderson was killed by a car as he pulled out of a gas station aboard one of his new machines. But the factory stayed alive thanks to the defection back to Ace of Excelsior's chief engineer and old partner of Henderson, the stalwart Arthur Lemon.

To grab public attention, Lemon built a lightweight bike and achieved a recognized world record of 129mph. But the company would never be financially stable and eventually sold out to Indian, who began a fresh chapter in the story of Henderson's four. The Indian Four would be the flagship of the marque right up until 1942 when production ended.

Indian

Indian Chief

One of the most famous and universally desired classic American motorcycles ever built is the Indian Chief. The bike was christened back in 1922 and 'Chief' remained a model name until the company's demise in 1953.

The Chief was always the heavyweight in the Indian line-up. It was a machine favored by touring fans, by police forces and sidecar enthusiasts, not forgetting the various bands of roughnecks who stripped their Chiefs to bare essentials and took them on the war-path through sleepy Californian towns in the late 1950s.

The first Chief was fundamentally an enlarged version of the Scout, Indian's popular middleweight. It was a side-valve 42-degree V-twin (contrasted with Harley-Davidson's 45-degree twins) and was of 61 cubic inches (1000cc) capacity. In 1925 a 74-cubic-inch (1200cc) model was added to the range. The Chief had a magneto and generator to power the lights and brighten the sparks. Like the Scout (and unlike the Harley) it was of unit construction with helical gear primary drive and a three-speed gearbox. It was good for a very impressive (and probably very scary) 90mph. In 1922 the Chief went on sale in the USA for $435.

The Chief stayed in production until 1953 and in its final guise was perhaps the most handsome American motorcycle ever built, with it's bright-red paintwork, deeply valanced fenders and stylish detail (a running light in the shape of the head of an Indian chief graced the front fender). The model ended its life with a specification which boasted an 80-cubic-inch displacement, a twist-grip throttle on the left bar, and a hand-controlled stick-shift on the right side of the bike.

Below: *A late Indian Chief with telescopic forks and plunger rear suspension. The Chief, which enjoyed a 31-year production run, was introduced in 1922 and sold for $435.*

Above: *One of the finer details which separated the Indian Chief from the cowboy rivals: the distinctive Indian brave running light which graced the front fender.*

Harley-Davidson 61J

The Harley-Davidson 61J was, to the motorcycling enthusiasts of the 1920s, the definitive Harley-Davidson. It was rugged, reliable and had all the pull and grunt the big 61-cubic-inch V-twin promised. Its history goes back to 1909 when Harley's first V-twin, a pocket-valve 61-cubic-inch model was designed by the simple addition of a second cylinder to their popular single 'Silent Gray Fellow.' It set the trend for Harley production to the present day: big-inch 45-degree V-twin configuration became the trademark of Harley and American motorcycling.

In 1914, at the outbreak of war in Europe, the first serious revision of Harley's V-twin took place, designated the 'J' model; it would be the marque's flagship for 15 years. It featured a three-speed gearbox and a multi-plate clutch – both innovations for Harley. In 1921 a 74-cubic-inch version was offered to keen tourers and sidecar fans. In 1917 the J was updated, with a more powerful and strengthened engine; in 1919 the Model 20 offered an option of electrical equipment with generator, lighting, battery and horn as an alternative to the simple magneto and acetylene lighting version, the JF.

The basic J design remained unchanged until 1924 and the bike was Harley's mainstay through the period. A red-hot two-cam version for racing (and later, road use) was designed in 1924 by Arthur Constantine and Bill Ottaway and enjoyed limited production until 1929. It was designated the JH and is today a rare classic.

The J model was subtly updated in 1924, the conservative Arthur Davidson and Bill Harley not wishing to risk, by trying anything too radical, the great goodwill the model had accrued.

Right: *A 1919 Harley-Davidson 61 'races' through the haze of exhaust fumes.*

Below: *This Harley 61 is stripped for racing. It was Harley's flagship model for 15 years.*

But in 1929 Harley-Davidson felt a change was needed – a new model to replace the J. It involved updating the motor's old pocket-valve design with a side-valve arrangement. Their new bike was a big heavy monster with disastrous teething problems. Harley-Davidson had got it all wrong; after the wonderful J, the V-model was a sad disappointment. The heyday of the J was over but in history it remains a classic.

Harley-Davidson 21 'Peashooter'

The single-cylinder 'Peashooter' Harley-Davidson, like many 'classics,' began life in 1925 as

something of a flop. It was a 21-cubic-inch (350cc) side-valve vertical single, capable of a meager 45mph. It was succeeded by an overhead valve model which gave it another 20mph, but the flat-track racing craze of the era inspired the AA model, which was quickly nicknamed the Peashooter.

The bike would dominate the flat-track and hill-climb events in Britain, Australia and on the domestic racing scene beyond its production run which ended in 1934. Joe Petrali, the man who still holds the ocean-sands speed record for his 1936 run at Daytona beach on a 61 OHV prototype Harley, was the hero of the day, beating the Indians, the Excelsiors and any other upstart marque which dared to challenge. In 1935 he not only won the National Hillclimbing Championship on a Peashooter, but achieved the Grand Slam of the day, winning all 13 AMA (American Motorcycle Association) National Championship rounds of the American dirt track racing series. With little opposition from other American marques, the advent of the powerful English JAP engine signaled the end of the Peashooter's run of success.

The road bike, ironically, was never a popular Harley model simply because it was a single-cylinder machine. Although it reveled in its sports guises, the bike's light weight and lack of low-rev pulling power failed to impress the American public: the enduring attitude of American bikers – if it ain't a big twin, it ain't a motorcycle – was formed right back in the 1930s and continues its hold to this day.

Above: *In its day the Peashooter was in fact a big gun on the race tracks of America. It dominated dirt-track events, hill climbs and the 21-cubic-inch class for track racing.*

Excelsior Super X

Excelsior was, along with Harley-Davidson and Indian, one of the three great American manufacturers of the early days of motorcycling. Run by the acquisitive Ignatz Schwinn, the marque's history follows similar lines to the other two factories', initially building simple single-cylinder bikes which naturally developed into V-twins by the second decade.

Schwinn was never averse to buying ideas off others and poaching staff from rivals for his factory and he seized the opportunity to buy the Henderson design from William Henderson. Another high-profile character in the American motorcycle world of the time was the innovative and brilliant engineer Arthur Constantine. Without official sanction and in direct conflict with the wishes of his bosses, Bill Harley and Arthur Davidson, in 1924 he presented a thorough and radical redesign of the venerable J model Harley. Accused of wasting company time and resources, he resigned.

Excelsior snapped him up along with his exciting design for a V-twin, which the churlish Arthur Davidson had spurned out of hand, to his everlasting regret. Arthur Constantine's new bike, catalogued the Excelsior Super X, set the style for a whole new class of bike – the 45-cubic-inch (750cc) 'middleweight.' And to Schwinn's eternal satisfaction, the research and development costs were all borne by the marque's main rival, Harley-Davidson.

The 45 was based on Harley's successful 61J model – a pocket-valve 45-degree V-twin, but it was a much streamlined machine, lower and more compact. It was light and fast, the new engine producing a healthy output for the relatively small capacity. Although there were many chassis innovations, too, Schwinn chose to use existing Excelsior chassis parts to house the motor. It filled a gap in the market and was an immediate hit. It continued in production until 1931, with a bored-out 61-cubic-inch version offered just prior to the marque's demise.

Schwinn was quick to close down his motorcycle operations in 1931 in the face of worldwide depression. Being a man who liked to swim with the stream, he cut his losses – unlike the bosses of Harley-Davidson and Indian, who struggled on through the Depression years, determined to preserve their businesses. Who knows what frustration Arthur Davidson felt on the launch and initial flop of Harley-Davidson's first 45-cubic-inch model in 1929. Ironically, however, it was the 45 which would help keep them afloat through the 1930s and into the war years, once the creases were ironed out.

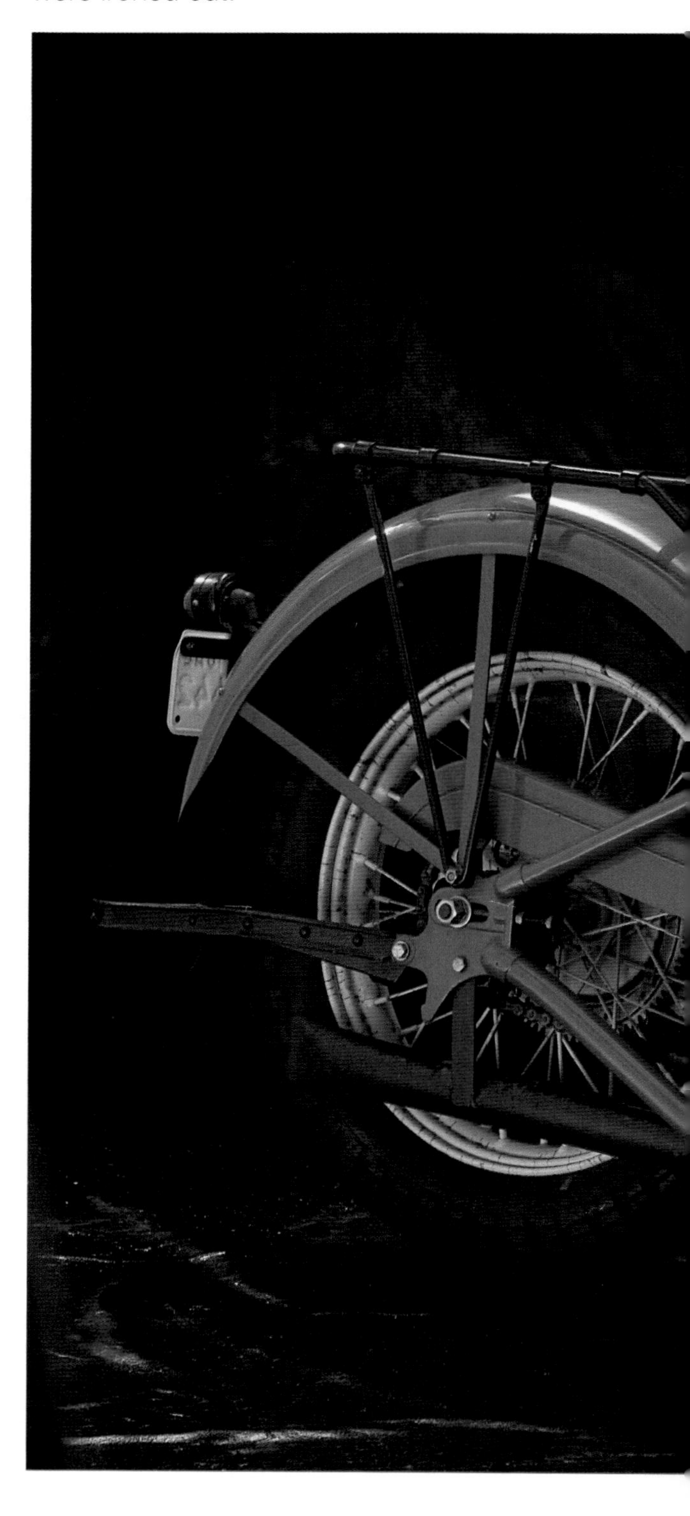

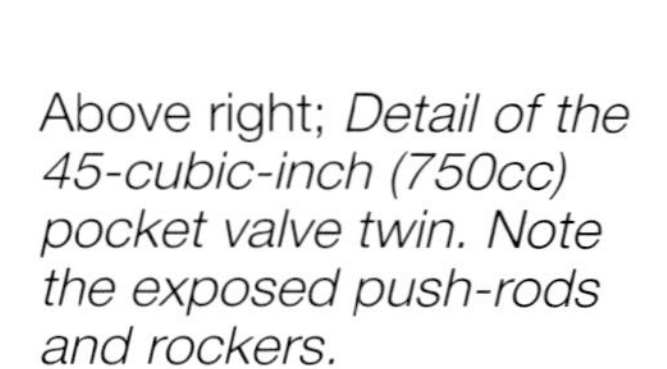

Above right; *Detail of the 45-cubic-inch (750cc) pocket valve twin. Note the exposed push-rods and rockers.*

Above far right: *Super X front end showing electric light and horn. Note the maker's emblem on the steering head.*

Right: *The Excelsior Super X was the front runner of an exciting middleweight class of the 1920s. The 45-cubic-inch (750cc) bike took the public by storm – ironically it was based on the design of a Harley-Davidson employee!*

SUPER
X

Above: *A Knucklehead of 1939, showing (above) its pair of shiny 'knuckles' – the distinctive rocker covers which gave the bike its nickname.*

Right: *The Knucklehead was Harley's first production bike with overhead valves, which contributed to a doubling of horsepower from its predecessor.*

Above right: *The streamlined prototype which reached 136mph to claim the world speed record in 1936.*

Harley-Davidson 61E 'Knucklehead'

It is a great event in motorcycling history when an all-new Harley-Davidson model is launched. The term all-new is of course relative, as Harley haven't changed their basic engine configuration for over 80 years. However there have been watershed models along the way which denote new chapters in the marque's history. The 61E 'Knucklehead' is just such a model.

The big difference boasted by this model was the use of overhead valves as opposed to pocket valves and side valves, the first American motorcycle to feature such an arrangement. The knobbly rocker covers gave the bike its unofficial monicker – the Knucklehead. The bike was a new bike from the ground upwards, bearing no relation, but for the cylinder configuration, to its predecessor, the VL. The new push-rod operated valve arrangement *doubled* the bike's output to 36hp. Capacity was 61-cubic inches (1000cc) and compression ratio was 6:1. It featured dry-sump lubrication (as would all new models thereafter), a wide-ratio four-speed gearbox and a massive multiplate clutch. Top speed was over 90mph – the fastest Harley to date.

Joe Petrali, Harley factory racer and co-design engineer on the project, set out to prove exactly that by running a stripped bike at Daytona Beach on 13 March 1936. He scored a record 136mph run – after removing the streamlining which caused the bike to take off at over 100mph! The record attempt (which still stands as an ocean-sands speed record, principally because no one has ever bothered to try to beat it) was something of a diversionary tactic to draw attention away from the model's teething problems of valve-gear failure and oil leaks.

Early niggles aside, the 61 Knucklehead remained the backbone of Harley heavyweight production right through until 1947 in both 61- and 74-cubic-inch form. It's been a collector's piece and a customizer's dream ever since. In 1948 the Panhead with its alloy heads and hydraulic lifters powered the big Harleys, opening a new chapter in Harley history.

HARLEY-DAVIDSON

Above and right: *The fastest, sweetest handling and most perfectly finished machine of its era: the Brough Superior SS100. Its high standard of performance and ludricrously expensive price-tag made it a status symbol of the rich and the noble – a strange achievement for a mode of transport considered, at the time, a plebian preserve.*

Brough Superior SS100

'The Rolls-Royce of Motorcycles,' as George Brough described his own product, is one of the most desired motorcycles of all time. It had speed (100+mph), exclusivity (only around 3000 were ever made in 20 years of production), romance (it was the great love of Lawrence of Arabia – and it cost him his life) and like a Rolls-Royce had a prohibitively high purchase price (£170) which confined it to the pockets of the very rich.

George Brough's bikes were his own design and style, but not entirely his own manufacture. He borrowed parts from many manufacturers, parts he considered the best, including the engine which was the huge and powerful JAP and later the Matchless V-twin.

The SS100 used the JAP engine from 1925 to 1935 and ran with Matchless 990cc motors until 1939. The JAP was the biggest engine in production at the time – the '90-bore' (90mm bore × 77mm stroke) and featured vertical overhead valves and a three-speed Sturmey Archer gearbox. In 1933 it was redesigned to 996cc, with a bore and stroke of 80mm × 99mm, giving 74hp at 6200rpm. It had twin magnetos, dry-sump lubrication, twin carbs and a four-speed transmission. Primary drive was by chain.

The Matchless motor fitted after 1935 was high-tech indeed, with hairpin valve springs, crank and mainshaft running phosphor-bronze bearings and a square bore and stroke of 85.5 × 85.5mm. A speed of 100mph was no problem.

The Brough's handling qualities, like all else about the marque, were legendary; at a time when speed wobbles were commonplace with simple leading-link sprung forks, Brough advertised: 'Hands off at 100mph!' Achieving 100mph was impressive enough!

At the Haydn Road Brough Superior factory in Nottingham, England, the Brough mechanics would strip down all supplied components (including engines of course), put right any engineering or design faults they came across, and rebuild them with the meticulous care of a Swiss watchmaker. Woe betide any supplier who sent faulty goods to George Brough's factory.

The result was a reliable, powerful and mechanically quiet motor finished to a standard that gave Rolls-Royce no cause for complaint

when their name was associated with Brough in advertising copy. The tank was hand-built from scratch using many sections, beaten to perfection and soldered together and polished to a mirror finish; it was the bike's crowning glory.

'I told myself that what was wanted was a big solo machine made up to an ideal and not down to a price: that it must have everything of the best in its manufacture . . .' George Brough wrote for *Motorcycling* in 1924. He insisted there was a niche for an 'ultra-luxurious solo' at a time of depression and falling prices. He was right – it was a small market and he exploited it.

The story of Brough ended as it began – with a dream, the Golden Dream. It was George's continuing search for the ultimate motorcycle, a flat-four – vibrationless, compact and powerful. Its realization was beyond Brough's resources, especially as war was spreading through Europe. George Brough found it increasingly difficult to support a business geared to such a luxury item. More pressing matters were at hand. His precision-engineering business was turned over to the war effort and his magic spell on the world of motorcycling was broken.

Above: *The first production model of the Triumph Speed Twin appeared with girder forks and rigid rear end. It was so stable – cornering or on the straight – that it was capable of out-running any opposition.*

Far right: *The classic parallel twin-engine configuration set the style of motorcycling for the next 20 years. The first model was 498cc and was capable of over 105mph – it was the bike which spawned a generation of 'ton-up boys.'*

Triumph Speed Twin 500

In July 1937 a motorcycle was launched which had more impact on world motorcycling than almost any model before – the Triumph Speed Twin. It was the first project of motorcycle engineering genius, Edward Turner, following his appointment at Triumph a year earlier. It was Turner who proved his talent in designing the Ariel Square Four.

The Speed Twin was the father of a new generation of motorcycles. It mimicked a single in that its parallel pistons rose and fell together in an even firing sequence and was surprisingly smooth. Almost overnight its simplicity and economy converted a motorcycling public from the cult of the single cylinder to the power and performance of the multi. On its launch in 1937 its price was £77.15s.

The Speed Twin's capacity was 498cc and it had a bore and stroke of 63 × 80mm. It used overhead valves operated by pushrods set between the two cylinders, with cams driven by gear wheels in front and behind the crankcase mouth. It had a four-speed gearbox, alloy crankcases and a cast-iron block. Its top speed was quoted in the press of the time as 107mph and the quarter-mile was achieved at 74mph from a standing start. The Speed Twin was a rocket.

The chassis was conventional for the time and similar to the single-cylinder Tiger 90 frame. *The Motor Cycle* noted at the time that the engine weighed less than the single and that the weight distribution of the two bikes was identical. The chassis used a rigid rear end, the sprung seat providing cushioning, and girder forks. Handling was said to be as light as on a 250cc bike and so stable that it would fool the rider into taking corners at higher speeds than usual. The bike was a head-down no-nonsense scratcher and the staid prose of the day wasn't enough to disguise the bike's wicked personality.

The Triumph factory was bombed flat by the Luftwaffe early in the war. Without the Speed Twin and its instant reputation as a tire-smoking road rocket, who knows whether the factory would have recovered, as it did so bravely after the war. The basic design had set a new worldwide industry standard and stayed in production until the Triumph factory closed forever in 1983.

TRIUMPH
MINIMUM OIL LEVEL

Above: *The Vincent Black Shadow had a chassis to match its big-bruiser engine. An innovative cantilever rear-suspension system and cleverly crafted 'Girdraulic' forks kept the bike rock steady up to speeds of 125mph and beyond.*

Vincent Black Shadow

If there's one thing that will always sell a motorcycle, it's speed. HRD Vincent motorcycles have a notoriety which still carries credibility even in the modern era of 170mph superbikes. A Vincent Black Shadow was the first motorcycle (unsupercharged) to travel at 150mph, and it reached this speed as long ago as 1948. Admittedly its rider, Rollie Free, made the run wearing nothing but a swim suit and a pair of running shoes – the drag from his leathers and boots would have robbed him of a couple of vital mph! Although that machine was lightly tuned, the standard Shadow could manage 125mph – if the rider could keep control on the rough roads of the day.

Phil Vincent was the man behind the marque. The HRD part of the name came from an established make (Howard Raymond Davies was its original proprietor) which Vincent took over. He felt the HRD name would lend his new bike some credibility. As it turned out, his bikes needed no such support.

His super torquey engines were V-twins, set at 50 degrees with overhead valves. Few bikes of the prewar days had alloy barrels and cylinder heads, and certainly no twin was as compact as the Vincent. The first V-twin was the Rapide, built in 1936, but the marque's greater fame came after the war with the launch of the Black Shadow in 1948. The engine's capacity was 998cc and compression was set at 7.3:1. Horsepower

topped 55 at 5700rpm. *The Motor Cycle* recorded 87mph in second gear and 110mph in third. They couldn't find a road long and smooth enough to achieve a flat-out top speed in fourth.

The bike was every schoolboy's dream – and every motorcyclist's too. Retail price for the Shadow in 1948 was £315 - a year's wages to most. The bike had a fearsomely fast motor, but also, thankfully, the chassis to contain it. Phil Vincent was years ahead of his time in chassis design: at a time when rigid rear ends were just being superseded by the swingarm, he incorporated a cantilever rear end which would emerge again on Japanese bikes some 30 years later. He shunned telescopic forks (which were just becoming popular) in favor of his far superior (at the time) Girdraulic fork. His bike needed forks with great lateral rigidity, which girder forks offered. But he used forged light-alloy blades of tapered oval section for lightness and strength. The trail was adjustable (for solo or sidecar use) and had effective damping with two telescopic spring dampers and a 'two-way' hydraulic damper unit which stopped sudden bottoming- or topping-out.

Vincent's glorious years were sadly short-lived. At the height of its success the company was in fact bankrupt and went into liquidation in 1949. Clever accounting kept Vincent producing bikes until 1955, but the company would never again reach the peak of excellence achieved in 1948 with the Black Shadow.

Above: *The enormous 150mph Smiths clock served as a useful windshield! The figures told every inquisitive schoolboy all he needed to know. The alloy knob is a steering damper – a standard fitment and a necessity on a high-speed machine traveling on low-speed roads.*

Harley-Davidson 45

The Harley 45 engine enjoyed a production run which outstripped any other motorcycle engine ever to have existed. It was in production, largely unchanged, at the Milwaukee factory for 45 years!

The 45 was a 750cc side-valve 45-degree V-twin which was built to rival the Indian 101 Scout and the Excelsior Super X. The engine was the basis for the famous KR750 racers as well as the powerplant for the three-wheeler delivery vehicle, the Servi-car, which went out of production in 1974 after 42 years.

The first 45, launched in 1929, was a total flop. It was intended to be a nimble middleweight but could barely achieve 55mph. It caused rebellion among Harley dealers who were embarrassed to offer the bike as a serious alternative to the competition – the Excelsior Super X could hit 90mph! Despite a poor start, the bike limped on in the Harley range. It was updated in 1932, and was

Above: *An ex-army Harley 45, de-mobbed and dressed for a new and more glamorous postwar life.*

Right: *A 1942 WLA, ready for battle. It was one of 88,000 machines which Harley-Davidson counted out of the factory as Milwaukee's contribution to the war effort.*

given a further boost by the discontinuation of the Indian Scout. That same year the motors began their run in the Servi-car.

By 1941 the 45 was an established favorite, a workhorse and a trustworthy, if unremarkable, middleweight. But it was on the threshold of something big, an event which sealed its destiny to become a classic – World War II. As the world's supply of motorcycles was being severely curtailed by the relentless activity of the German airforce over the industrial heartland of England, new demand for the 45 came from the Allies and from the US Defense Department. The 45 was about to become the soldier's friend.

By the war's end Harley had sent 88,000 45s to war, mostly the low-compression, sluggish but unburstable WLA model. As Harley didn't appear to anticipate the end of the war, they were left with a huge surplus. None went to waste of course, as they were easily converted to civilian spec for the thousands of eager and freshly demobbed GIs.

Above: *Harley's heavyweight tourer of 1949 was given hydraulic front suspension to soften the bumpy ride to an easy glide. The model also saw the introduction of the 'Panhead' motor.*

Harley-Davidson Hydra Glide

When Harley-Davidson fitted telescopic forks to their big tourer, the 74-cubic-inch (1200cc) FL model, in 1949, they turned their back on 40 years of leading link tradition. The Harley fraternity was outraged. Newfangled it may have seemed, but it was the way of the future. And it worked too, giving the big tourer a smooth ride. The sales literature capitalized on the 'new' hydraulic/spring suspension system (Triumph had three years of telescopics behind them already), dubbing the new machine the 'Hydra Glide.'

The Hydra Glide marked the end of the Knucklehead's reign as top hauler; it was fitted with the new Panhead motor ('pan' because of the shape of the rocker cover). 'New' is always a relative term with Harley, who have never been known for their innovative approach to design. The Panhead was a slight redesign of the Knuckle, aimed at curing the odd top-end glitches of the latter. The head was now cast in aluminum rather than iron and Harley introduced hydraulic valve lifters, which still feature on Harleys today, and which maintain precise valve actuation (Harley barrels and heads tend to grow a great deal between cold and hot).

The Hydra Glide retained its solid rear end – shock absorbers were a few years away yet – the bumps being taken care of by the sprung seat post and the huge balloon tires. Its status as a classic was reinforced in 1985 with the introduction of the 'Softail' in the modern Harley line-up. It has all the looks of the Hydra Glide, but modern behinds are spared a beating through use of two hidden shocks, laid horizontally under the engine.

Harley-Davidson 883 Sportster

The Harley Sportster 55, launched in 1957, opened up a whole new sporting road for Harley-Davidson enthusiasts. The motorcycling world had received a potent adrenalin injection by Triumph's Speed Twin variants, and this was Milwaukee's answer.

The bike was based on the K-model 45, but bored out to 55-cubic-inches and given overhead valves. The motor, housed in a spartan (460lb) sporting chassis, had a 40hp punch delivered at a busy 5500rpm. It was the bike for cruising the main drag, for holeshotting the hot-shots on the green light and out-hauling the British invaders with a 100mph top-end speed. The chassis was the established duplex frame, running swingarm suspension at the rear and telescopic forks at the front, neither of which, incidently, added to the bike's performance.

The model began a run of Sportster production at 883cc, which continues today, over 30 years on, with the engine's look, format and capacity largely unchanged.

Above and right: *From 1957 to the present day, the 55-cubic-inch (883cc) Sportster has underpinned the Harley-Davidson range as the entry-level hog. The machine pictured here is a 1987 model, with four-speed gearbox and one-speed attitude.*

Above: *Triumph took their Speed Twin another step forward in 1959 with the Bonneville 650. The 'Bonnie' was fast and reliable and came complete with a world-record-beating pedigree.*

Triumph Bonneville 120

When the Triumph Bonneville was launched in 1959, the bike's name was no mystery to the motorcycling world. Three years earlier at Bonneville Salt Flats, Utah, a streamliner with a Triumph Thunderbird-based engine, running on nitromethane, tripped the lights at a world-record-shattering 214mph.

The machine was being sold entirely on its speed potential and that, as with its predecessor, the Tiger 110, was contained in its name: Bonneville 120 signified 120mph! In 1959 the bike made its appearance at London's Earl's Court motorcycle show without a designation; the name came after.

What impressed people at the time was its high performance and rugged specification. The crank was now a one-piece forged item, capable of handling the kind of horsepower produced by the unit construction 649cc OHV parallel twin. The engine featured a 'splayed' double intake on the cylinder head, which used two Amal Monobloc carburetors and in tests was producing nigh-on 50hp. The prototype which weighed just 404lb, achieved 128mph at the track. Production bikes would, in fact, do well to get over 110mph, but the potential was established.

The first Bonneville had a single downtube frame, replaced the next year by a duplex cradle frame which would be the design for the bike until the last Bonneville was built in 1987. The 1960 model was summed up in the press reports as a 'roadburner with sparkling performance, tractable power and super brakes and roadholding.'

In 1987, in its final incarnation, the bike was a 750, using fundamentally the same motor as was unveiled in 1959. It had more torque, but top speed remained around the 115mph mark. It still leaked oil too. Compared with the Japanese machinery of 1987, though, *The Motor Cycle*'s opinion of it in 1960 no longer quite held true. As a machine of character, history and style, the Bonneville, even today, has few peers.

Harley-
Davidson
Classic

Harley-Davidson Electra Glide

The Harley-Davidson Electra Glide is one of the most famous motorcycles of all time. Ask any enthusiast in the Western World to give the name of a classic motorcycle and that's the one they'd mention. Why? Well, it helps to have a evocative name such as 'Electra Glide' coupled with what is clearly the most well-known and longest established motorcycle marque in existence. It also helps to have a movie named after it; viz – *Electra Glide in Blue* (1973) starring Robert Blake.

Since 1965 the Electra Glide has been acknowledged as the heaviest of heavyweight tourers, a pinnacle of lazy touring luxury. Harley-Davidson themselves crowned the bike 'The King of the Highway,' a title which perhaps only the Honda Gold Wing has had the temerity to challenge thus far.

The 'Electra Glide' monicker was derived quite simply. The bike's predecessor was the FLH Duo Glide, so named after its 'innovative' swingarm and suspension units which replaced the previously rigid rear end. It was the electric starter version of the Duo. The motor was, of course, the obligatory 45-degree V-twin, with separate 4-speed transmission and a 1200cc displacement. In 1966 the 'Panhead' motor was updated to 'Shovelhead' specification and was producing 60hp, a reasonable enough measure of grunt, until you consider the bike's 780lb weight disadvantage. To this day the bike ranks as one of the heaviest around. Nevertheless, even at low speed this two-wheeled motorhome (modern models have stereo hi-fi as standard) handles like a bike half its weight.

The Electra Glide will lope along at 55, 70 or 90mph and never feel strained. At 55mph the engine is ticking over smoothly at a mere 2600rpm! No bike matches the Electra Glide for its ability to relax the rider rather than raise the blood pressure.

Left and below: *The ultimate in touring luxury, combined with charisma and style. The Harley-Davidson Electra Glide is an unrivalled experience in the world of motorcycling. The model was launched in 1966 and continues in the range today, same looks and image but a much refined ride. The model depicted is a 1989 FLHTC.*

Far right: *Launched in 1966, the Honda CB450 was the West's first taste of Japanese precision engineering and impressive technology.*

Right and right below: *Three years later, the world gasped in awe as the new standard for motorcycling was set by the Honda CB750 Four. The bike rewrote the book in terms of performance, reliability and engineering standards.*

Overleaf: *The author rides the MV Agusta 750S, the first road version of the legendary Grand Prix 500 racing bike. The model carries with it the history and charisma of 37 World Championship titles.*

Honda CB 450

The Black Bomber. Honda gave it the name with no explanation (other than the tank was black and chrome) and began punting it out to the world in 1966. It was a 444cc parallel four-stroke twin, with double overhead camshafts, a feature reserved until then for quixotic racing specials. The system was effective if rather elaborate. The CB450 featured constant velocity carbs, today an industry standard, then an innovation. Gearing was low, aimed at keeping the revs high where the bike's peak 43hp lay. Another trend being set, for the Japanese at least, was for reliability and oil-tight cases. The CB450 promised thrilling performance for its specification and £560 asking price.

It certainly had the performance, likened by owners to that of 650cc British bikes of the day. It had around 105mph top speed, but startling acceleration up to it – startling for any British twins it may have been pitted against anyway. The Black Bomber had high-rev power and low-speed torque; it handled extremely well, but for a reservation over the rear shock absorbers. It was clean-running and started electrically, on the button – another new trend for 1966.

What Honda had created with the CB450 was not so much a highly modern roadburner as an all-rounder which set new standards for the industry to ponder and try to emulate. In concept it was the precursor of the CB750 four; in practice it was an unrivaled all-purpose motorcycle.

Honda CB750

With the introduction of the Honda CB750 at the Tokyo Show in October 1968 came the introduction of a new term to the English language: superbike. The bike was so unlike anything the press or public had seen before, a new word was needed to describe the father of a new family of extreme performers.

The air-cooled 736cc (bore × stroke 61 × 63mm) parallel four-cylinder motor had single overhead camshafts, five-speed gearbox and made 67hp at 8000rpm. Top speed was an effortless 120mph and it covered a quarter-mile from standstill in just 13.5 seconds. Its performance could be matched by other bikes around at

the time, such as Triumph's Trident and Kawasaki's Mach III 500, but it was the super-smooth turbine-like generation of the power which really set it apart. That and the quality of engineering which made the bike one of the most hard-wearing and reliable on the market.

The chassis to hold such a powerful hunk was a steel duplex cradle, suspended by gaitered telescopic forks at the front, and twin shock absorbers at the rear. It gave a firm but sporty ride and could be pulled up quickly and with fine control by the 290mm (11.5in) front hydraulic disk-brake. The bike set new standards for an industry which was advancing in leaps and bounds anyway, driven on by tireless Japanese engineering enthusiasm.

In its first acquaintance with the bike, *Cycle World* magazine commented: 'If you had the finest of all production machines, a two-wheeled answer to Ferrari-Porsche-Lamborghini, you would be riding a Honda 750cc four cylinder.'

CB 450

HONDA

AGUSTA

750S

Above: *Pure Italian style, born of racetrack experience. Later models would wear full fairings, but the 750S was too proud of its beautiful engine to contemplate such modesty.*

MV Agusta 750S

The Italians have always taken their motorcycles very seriously and, in typical Latin style, very passionately. Count Domenico Agusta was just such a man, who refused to compromise his motorcycle design for the sake of money. The result was an unparalleled 37 World Championships for the marque and immortality in motorcycle history.

The sporting road bikes MV Agusta produced were as uncompromising as his racing machines. But the Count always resisted building a road-going race replica for his clamoring fans. In 1966 he built a 600cc tourer, a rather dull-looking machine next to the GP bikes which so fired the blood of race fans. The tourer, rather predictably, was a failure, but then the Count gave in and launched the 750S – which was exactly what the world had been waiting for.

In 1972 the 750S was well ahead of the Japanese. The nearest to its technological standard was the Honda 750-4. Its specification was that of the 500cc GP bike; it was an air-cooled four-cylinder motor, with oversquare bore and stroke (65 × 56mm), it ran with double overhead cams driven by gears and the cases were of sandcast aluminum. It had four carbs and four long, swooping bright chrome exhausts. Peak power was at 7900rpm, a respectable (but under-developed) 65hp. It was also shaft drive, something the Count used more for tradition than mechanical efficiency. It inspired MV tuner Arturo Magni to design a parallelogram double swing-arm in an effort to counteract torque reaction.

MV Agusta went bust. Their uncompromising roadbikes were most of the reason, being simply too expensive for mass-production and too expensive for most potential buyers. In 1972 a Honda CB750 cost just £900 while by comparison an MV Agusta 750S cost a heart-stopping £2175. The most desirable bike of its time was also the most unobtainable.

Norton Commando

The Norton Commando was the swan-song of the British motorcycle industry – the last great British roadburner. It was launched in 1968 and marked a new, if limited, direction for Norton. The design left behind a tradition of vertical singles and twins housed in the superb Norton Featherbed frame. The first Commando's capacity was 750cc (73 × 89mm), a simple pushrod twin with inclined forward cylinders, its pistons rising and falling together in a primarily unbalanced configuration. Nevertheless, the engine produced enough power – 56hp – to push it to over 120mph.

The reason for the Commando's success is partly the reason why its life was to be so limited. While in Japan the industry was looking at engineering means of producing power without vibration. Norton circumvented the problem by building a frame – the Isolastic frame – which put rubber in between the buzzing motor and sensitive flesh and bone. Honda's 750 four would show the world the true way, just two years later.

The rubber-mounting system allowed the powerful engine to be used – to be kept at high revs without the rider losing any fillings from his teeth. It was a reliable motor and as a road-test report said at the time, 'is probably more oil-tight' than the bike it replaced, the Norton Atlas. The Japanese were soon to demonstrate that the phrase 'oil-tight' is a finite term.

The Commando got better-looking with age. Originally seen with a straight, streamlined tail unit reminiscent of a prop from Flash Gordon, plus cow-horn bars, it appeared a year later with high-level exhaust pipes and chrome galore. 1971 saw something of relapse in styling with the 750 Hi-rider, designed in the *Easy Rider* era with chopper handlebars and a banana seat. But by the end of its production life in 1975, the bike was an 850, painted black with gold highlights, a nod to John Player, the cigarette manufacturer who sponsored the factory racebike. It was perhaps the last roadbike to have a 'down-for-up' gearlever arrangement.

Below: *A late model Commando 850 with electric starter. It was a bike capable of a true 120mph with fine handling to boot. In performance it held its own against the Japanese fours of the early 1970s, but it was at the end of its development, and they had just begun.*

Above and right: *The original rice-burner. Kawasaki's Mach III set the factory on a notorious course which they still hold today – as builders of the world's fastest production motorcycles.*

Kawasaki 500 Mach III

The name Kawasaki means one thing to a biker: acceleration and crazy speed. The craziest was certainly Kawasaki's first attempt at a tire-smoker: the H1 500, aka the Mach III, the fastest bike and upstart rebel of the decade.

The engine was like nothing seen before: a parallel air-cooled two-stroke triple. It produced 60hp and ran in a 400lb chassis. The power delivery was described in the press as anything from 'mind-blowing' to 'nerve-wracking.' It was an exciting motorcycle either way. It introduced the wheelie to the motorcycling public as an everyday facet of Mach III ownership. Flipping the bike at the traffic lights was a real danger, despite Kawasaki building another two inches into the wheelbase before production began. Those two inches still only took it to 51 inches. It was all the owner could do to keep the front tire and road in contact. Top speed was around 120mph, if the rider could hold on to the weaving beast; standing quarter-mile times of around 13 seconds were recorded.

Press comment of the day was ambivalent. With the Mach III they were presented with a bike which had a hinge in the middle, or so it seemed during spirited cornering, with an engine that drew analogies with explosive weapons. It was exciting, scary, above all fun, but, as one American journalist commented: 'I would say that some of our more impetuous lads are going to get into difficulties with this motorcycle.' And they did.

KAWASAKI
MACH
KAWASAKI

KAWASAKI

900
DOUBLE OVERHEAD CAMSHAFT
DOHC

Kawasaki Z1 900

In 1972 Kawasaki unleashed their four-cylinder 903cc monster on the world. The Big K introduced motorcyclists to a new height of performance, of acceleration and speed which had formerly been the preserve of the professional racer. The motor was a powerhouse with mechanical strength never before found on a motorcycle. It had an amazing top speed of 130mph; it could run a quarter-mile in 12 seconds from a standstill. The rush of speed as the double overhead cam spun with the motor through 6000 and beyond 9000rpm was a new and exhilarating experience for a public only just beginning to come to terms with the power and sophistication of the Honda 750 four. The Z1 was an overnight sensation.

It came on the market with distinctive high bars, a large sleek tank with side panels and tail unit which created a new look, and one which would be seen again and again on Japanese bikes.

The first Kawasaki 900 sported just one disk brake at the front and a drum rear. It wasn't enough. Nor was the rigidity of the conventional duplex cradle frame and relatively poor suspension. The motor was too much for the chassis technology of the time – fine in a straight line, but big trouble in the turns. As a result, the brutish motor found its way into countless special aftermarket chassis and of course, on to the racetracks and dragstrips.

All pictures: *Kawasaki's Z1 900 took the sophistication of the Honda 750-4 and added brutal energy. Its double overhead cam four-cylinder motor redefined the word 'fast' for bikers the world over with a top speed of 130mph.*

SUZUKI
WATER COOLED

Suzuki GT750

After Kawasaki's vicious 750H2 triple-cylinder two-stroke, the Suzuki GT750 was nothing more than a pussycat. Affectionately known as the 'Kettle' it developed the basic idea away from stomach-churning, wrist-wrenching performance towards user-friendliness and refinement. The motor, still a 750 two-stroke triple, was water-cooled and soft-tuned. It was held in rubber mounts which isolated the rider from unpleasant tingles, and was mechanically quiet thanks to the water-jackets around the barrels. A speed of 105mph was the best to hope for and full-bore acceleration was uneventful thanks to an even spread of power.

The horsepower claimed at the time was 67, the same as a Honda CB750 which seemed to use its output to much greater effect. For the thrill-seekers whose appetites were whetted by the manic Kawasaki, the Suzuki could only bring disappointment. Even the Mach III 500 would leave the Suzuki looking sick. For those preferring a smooth and sophisticated alternative to the Honda four with similar sports and touring aptitudes, the bike was an attractive option – even if it did seem to promise more.

In fact, that was not an empty promise, as was subsequently proved by British racer Barry Sheene who campaigned a GT750-based racer in Formula 750 racing throughout the 1970s. His machine, at the hands of Suzuki factory tuners, turned from mild-mannered Kettle to a scalding 100hp, 170mph whistling missile which took him and Suzuki to international stardom.

Above and left: *Mild mannered, but full of two-stroke character, the under-tuned GT750 'Kettle' fell somewhere between the sports and touring category as a roadbike. In the hands of the racing boys, though, the engine was a championship winner with the likes of Britain's Barry Sheene on board.*

Above: *An early Wing, dressed and customized for long-haul posing.*

Above right: *The glamour of a gleaming Gold Wing Executive.*

Right: *The six-cylinder GL1500 – as far as the concept can be taken?*

Honda Gold Wing GL1000

The GL1000 was the founder of a dynasty of Gold Wing super-tourers, a series of machines which have achieved cult status the world over. It began in 1975 with a giant machine, which scorned all ideals of lightness and power-to-weight advantages, by mixing engineering sophistication with sheer bulk.

The bike in its first incarnation was unfaired, weighed 571lb dry and produced 80hp at a lazy 7000rpm. The engine configuration was a transverse flat-four SOHC with watercooling and shaft final drive. Its top speed was just over 120mph. Besides its out-and-out power, the almost vibration-free delivery of the thrust and the bike's overweight stability won fans in every country in which it appeared. It was a bike either loved or loathed. Those who loved it have built a worldwide affiliated organization which is unparalleled for an owners' club for one model.

The Gold Wing fanaticism has its strongest base in America but it is fair to say that owners throughout the world tend toward a certain stereotype – well-heeled, middle-aged and clean in every respect. The States were soon supporting a market of 25,000 bikes a year and the demand for aftermarket touring accessories mushroomed. So much so, in fact, that it wasn't long before Honda began dressing their fat mother-of-all-bikes to suit its fan club. The bike itself was being assembled in America and with each model change, the Gold Wing concept grew bigger and bigger. Production continued and reached a peak in 1990.

Undisputedly the ultimate touring machine ever built was the Honda GL1500 Gold Wing. It was an enormous beauty, weighing 800lb, powered by a 100hp flat-six engine. The bike, if it could be called that, was laden with gadgets and creature comforts ranging from hi-fi stereo to adjustable pillion footrests, the whole being built to a standard of quality more usually associated with luxury cars than with motorcycles.

HONDA

SHOEI

Above: *The most evocative of Italian sports bikes, the Ducati 900SS was built simply to be ridden fast, with no other consideration. Aesthetically, the skeletal machine had simple, functional beauty.*

Ducati 900SS

For some enthusiasts, all Ducati motorcycles are classics. The marque possesses a special romance and charisma in the world of sporting motorcycles. Of all their bikes perhaps the 900 Super Sport embodies all that is unique in a Ducati.

Based on the famous 750SS, the 900SS was produced in 1975 only as a short-run hand-built 'special.' Because its ugly sister, the 860 GT, bombed in the showrooms, the factory in desperation pumped all their resources into what was actually selling – the 900SS.

The 900SS was essentially a café racer, with half-fairing and macho 90-degree V-twin hanging shamelessly in the breeze. The model was predestined to be a success, as the 750SS, the bike it replaced, was instantly accorded classic status. The 900 version was bored out to 864cc (86mm × 74.4mm) and featured the famous Ducati desmodromic valve gear which was a mechanical method of positively shutting the valves as well as opening them. The system gives the motor better breathing and the possibility of a higher rev ceiling. The overhead cams were driven by bevel gear. The motor would rev comfortably to 7900rpm and had a compression ratio of 9.5:1.

Super Sports was no idle claim. The bike had a rich history of race success in 750 form, notably in the 1972 Imola 200, where it took first and second place against a horde of BSAs, Kawasakis, Hondas and Suzukis. In 1978 Mike Hailwood, a man who had won 76 Grands Prix in his career, returned to the Isle of Man after an 11-year 'retirement' from bikes to contest the TT. He won the Formula One event and lapped at a record-breaking 110mph lap of the 374-mile road circuit. His bike was a race-prepared 900SS.

Moto Guzzi Le Mans

In 1976 a push-rod operated OHV valve, air-cooled V-twin could still cut it as a hot bike. The Moto Guzzi Le Mans certainly did. The engine specification was basically that of a sedate tourer, but the Italians just can't let sleeping dogs lie. It ran with 10.2:1 compression ratio and produced 81hp at 7300rpm. Its top speed was around 135mph – about the same as the Kawasaki Z900! It was styled like a typical café racer: low bars, small headlamp cowl, a minimalist seat, and rear-sets, winning many over just by its sheer good looks.

The Le Mans made an impression mainly because it was rather unexpected that a lazy 844cc 90-degree V-twin could go so fast – 13.5 seconds over the quarter-mile. It revved, but still had the thumping pull of a big twin down in the low range. It was a shaft-driven bike, which in-

Left and below: *Arguably the finest Moto Guzzi ever built, the Le Mans 850. Although the shaft-drive V-twin motor seemed more suited to casual touring, in a high state of tune the power delivery of the 'Lemon' proved to be astounding.*

Overleaf: *The Ducati 900SS, stiffly sprung for high-speed stability, was lauded for its light handling and integrity in fast bends.*

volved the gentle throttle and clutch operation associated more with relaxed touring than with the violent stop-go extremes of a sports bike. Indeed it was a bike which needed a measured hand if it was to be used to the limit – some would say it was plain hard work.

The Le Mans boasted a safety feature which actually made the bike easier to ride – linked brakes. The bike ran with two cast-iron disks at the front, one of which was operated by the handbrake in the conventional manner, the other being activated with the footbrake in tandem with the rear brake – reassuring in the wet and in emergencies.

Guzzi built a whole series of Le Mans models, ending with the Le Mans V, the fifth variation on the theme, but it is generally agreed, as is so often the case, that they never surpassed the style of the very first Le Mans.

DESMO

900
SUPER SPORT
DUCATI

Laverda Jota

The Italians do not like to compromise. Ducati, Bimota, MV Agusta, are all testament to that. Laverda was a marque which liked its reputation for building big, fast muscle-bikes. And their bikes were just that. After success with a 750cc twin, the factory built what became known as the Beast of Breganze – the Laverda 1000 3C. Not many bikes in the mid-1970s had such a brutish specification: air-cooled, three cylinders, double overhead cams and a lumpy 180-degree crankshaft (later models went vibe-free with a 120-degree version). It was good for 125mph and was notoriously noisy.

Then the British importer at the time, Slater Bros, tuned one up and called it the Jota. The

Above: *The first BMW Gran Turismo machine – a fine handling and sporty bike with an unrivaled ability for long-distance endurance, thanks to a cleverly-designed fairing and a powerful, dependable motor.*

Opposite, both: *The Beast of Breganze, all torque and horsepower. The Italian Laverda Jota stood up to the blistering hi-tech machines from Japan as the fastest roadbike of the mid-1970s.*

meaning of the word is actually a three-beat Spanish flamenco dance, passionately performed of course. High-compression pistons, 38mm Dell'Orto carbs, open exhausts and more radical cam profiles gave the Jota so much extra grunt and top-end power, it shot straight ahead as the fastest production bike around – 140mph. It bellowed its dominance with an exhaust note so loud, everyone knew it was the boss.

The factory chiefs were impressed and weren't too proud to adopt the idea themselves and the Jota went into factory production, albeit a slightly detuned version of the Slater bike. The Jota was a hefty weapon to wield; the bike weighed a ton, it was tall, it vibrated, its clutch needed two hands to pull in, it had to be wrestled through bends with determination. It was loud and bad-mannered but as obvious as its red-painted tank the bike had that indefinable asset: charisma.

BMW R100RS

The BMW R100RS is the bike that bridged the gap between sports bikes and tourers, a bike built for sustained high speed either in a long, long straight line, or around twisty roads. It could get you across country – any country – faster than anything else the 1970s could offer. Ten years on and still in production, it was still the ultimate 500-mile-a-day bike despite BMW doing little to develop the model.

It was launched in the fall of 1976, a development of the R90S, and grabbed attention with its enormous sharp-nosed fairing which effectively gave complete rider protection without unduly compromising the aerodynamic efficiency of the bike. The waisted shape of the fairing had a down-thrust effect which kept the bike well in touch with the tarmac at high speeds. BMW went to great lengths to achieve this, using the wind tunnel of Pininfarina, the Italian automotive design company, to create the best shape.

The motor was the simple, fifty-year-old, horizontally opposed twin design. It was air-cooled, had two valves per cylinder, overhead cams operated by push-rods and shaft drive. Its top speed was just over 120mph. Its reliability is legendary. Until BMW designed their Paralever system of rear suspension, the bike suffered greatly from torque reaction of the shaft under acceleration and deceleration. The new system almost eliminated the rise and fall with the throttle.

The RS was widely used by the police in Europe as a traffic control-and-pursuit vehicle until the K-series models superseded it.

Harley-Davidson XR1000

It's not unusual for a bike generally considered by the experts to be a classic to endure life as a flop at the whim of public taste. The Harley XR1000 was just such a motorcycle. The bike was hyped in all the right ways: it was to be fast and punchy, based on the factory production racer and long-time winner, the XR750, and was promoted as such. It didn't quite turn out that way, though.

The idea was put into action by adapting the existing Sportster using XR750 racer cylinder heads. Renowned Harley tuner Jerry Branch was contracted to prepare the heads for the short-production-run machine. The intention was to sell the bike with an option to buy a Branch hot-up kit for those who wanted to race or simply leave black strips away from the traffic lights. What eventually went on sale had no more power than a standard Sportster. To compound the problem, Harley-Davidson were so impressed with initial dealer response to the idea that they hiked up the price by around $2000 from their initial retail price projection to $7000. The XR1000 died on the showroom floor.

So how did it become a classic? The Branch tuning kit had everything to do with it. Those XR owners who spent the extra $1000 on a tuning kit found Branch to be a man of his word; it nearly doubled the XR's output to 100hp. The model was discontinued in 1985 having earned a reputation as being over-priced and over-rated. Meanwhile, in backyards across America, speed-hungry bikers who had invested in the tuning kit were making exciting discoveries . . .

Harley-Davidson FXRS 1340

'There ain't no substitute for cubes' is a view to which Harley-Davidson and their fans have always subscribed but during the 1970s, many loyal H-D enthusiasts turned their backs on the marque in a search for something other than trashcan-sized pistons: reliability. It was a period of dire quality-control problems for Harley which inspired such sarcastic couplets as, 'Buy a Harley, buy the best; ride a mile and walk the rest!'

A management buy-out signaled a new beginning for Harley-Davidson and with it a new generation of motors: the Evolution engines. The first bike to carry the totally revised motor was the FXRS Sport Glide, a naked, low-riding simple big-inch hog. Harley folk didn't know what had hit

Right: *The XR1000, once scoffed at in dealer showrooms, now attracts crowds of admirers at bike shows everywhere, but its hidden potential was realized too late for commercial success.*

Left and below left: *The new Harley Evolution engine first seen in the 'Low Rider' FXRS would have been better named 'Revolution.' Its advent spelled a new era of success for Harley-Davidson.*

them. The bike was oil-tight – completely. The bike was reliable – totally. The bike didn't vibrate unpleasantly – a smooth Harley? A new rubber mounting system had taken care of the untoward tingles the primarily imbalanced engine usually transmitted to the extremities of the hapless Harley enthusiast.

The FXRS wasn't perfect by any means, but at least it was adequate in most departments: handling was acceptable (for a 600lb machine), engine performance punchy, if a little muffled (and nothing a poke up the exhaust pipe with a broom handle wouldn't cure); the brakes – well it's just as well the bike wasn't fast. The big 80-cube (1340cc) engine with its hopelessly old-fashioned separate five-speed gearbox would find itself, largely unchanged, at the heart of all Harley's models (Sportsters excluded) for the next decade. Indeed – why change a good thing?

The new engine and standards of engineering at Milwaukee and York, Pennsylvania were the basis for Harley-Davidson's renaissance from a struggling, near-bankrupt, out-moded and out-marketed business, to the major motorcycling success story of the 1980s. A restored reputation for reliability and clean running plus a clever capitalization on that intangible asset, Harley 'character,' have made the company stronger than it has ever been in its 90-year history.

Ducati 851 Superbike

The Ducati 851 Superbike boomed into life in 1987, a masterpiece of Italian engineering. Adhering to the factory tradition of building four-stroke 90-degree V-twins, the designer Massimo Bordi built a motor to take Ducati through the 1990s and secure the future of the classic Italian marque.

The 851 was built with these four aims in mind: compactness, light weight, high power and great torque. It easily fulfilled them all and in doing so challenged and trampled underfoot the Japanese opposition.

Bordi's engine was thoroughly modern: DOHC, water cooling, fuel injection, four valves per cylinder and, as he saw it, the key to the twins' amazing power, desmodromic valve gear. 'Desmo' means that the valves are positively shut as well as opened; the whole process is more precise than relying on a compressed spring to do the work. It gives better gas flow at low valve lift, and also a higher rev ceiling. It was a system which even Formula One car teams began to take seriously once its potential had been demonstrated.

The 851 was expensive for all its techno-wizardry, but proved reliable, and above all gave the power it promised: over 100hp for the road bike and up to 130hp for factory-prepared racers, which were also on sale to the public. With capacity raised to 888cc, the bike fulfilled its sporting raison d'être in 1990 and 1991, winning the World Superbike Championship, comprehensively beating its rivals Yamaha, Honda, Suzuki and Kawasaki to the laurels.

Right: *Ducati, the Italian factory, won the World Superbike Championship in 1990 and 1991 with the 851 Superbike in the face of the stiffest Japanese competition. Factory rider Raymond Roche was 1990 champion.*

Below and below right: *The stunning bright-red clothes of the Ducati 851 hide a complicated inner beauty. It has a tubular-steel space frame and the V-twin engine is a stressed chassis member*

SHARK
3

PURE SPORTS
FZ750
VBS

YAMAHA
5VALVE
FZ

Previous page and above: *Yamaha's FZ750, a tire-scorcher with a techno-marvel engine: five valves to each of its four cylinders; a canted forward block to give a downdraft effect into the carbs; it was the most compact 750-four ever made.*

Above right and right: *A sculpture of metal and glass fiber, the Bimota DB1 was more an Italian work of art than a motorcycle – but this was a bike to be ridden, not confined to a private collection.*

Yamaha FZ750

In 1985 Yamaha introduced a sports 750 which electrified the biking world: the Yamaha FZ750. It was the technical specification of the engine which promised so much and intrigued even the least technically minded. The engine was ostensibly a straightforward water-cooled, four-cylinder, four-stroke with double overhead cams – a familiar enough layout. What fascinated everyone, including those in the car world, was how and why Yamaha had built *five* valves per cylinder into their engine.

The reason was to create good gas flow and boost the power, especially in the bike's responsiveness and torque. The intake port was opened and closed with three valves rather than the then-conventional two-inlet, two-exhaust arrangement. The idea was that three valves gave a greater surface area for the inlet charge to get past. As the valves were smaller and lighter, the rev ceiling could be raised. High revs and high compression – 11.2:1 – made the motor 100hp strong.

But Yamaha didn't stop there with their fancy ideas about gas-flow. The cylinder block was canted forward from the horizontal by 35 degrees. The effect was to create a straight down-draft through the four carbs into the combustion chamber and out. Gravity could do its bit to help boost the bike's power too.

The motor's basic design has remained to this day virtually untouched in all of Yamaha's supersports machines. Indeed Bimota, the Italian sportsbike manufacturer, won the World Formula One Championship in 1987 with a fuel-injected FZ engine in their Bimota YB4. The success of the idea is clear to riders of FZ-based machines; high speed plus vivid acceleration equal adrenalin and thrills – the formula for success.

Bimota DB1

A Bimota is a motorcyclist's dream. To own one would mean two things: one, you're very rich and two, you have the most desirable sports motorcycle in the world. Bimota have an enviable reputation in motorcycling and a profile and presence as large as any manufacturer. More than that, the name carries with it a special charisma in modern motorcycling and perhaps analogous to that of the Brough Superior back in the 1930s. Yet the small company, based in Rimini, Italy, employs no more than 50 people.

The Bimota DB1 was a model which explains all. It is hand-built. The quality of its components was beyond reproach and its design not only makes it one of the best-handling bikes of its era, but also the most beautiful. The frame was of chrome-molybdenum steel, forming a triangulated trellis, using the engine – a Ducati 750 Pantah – as a stressed member. It was strong and light, the motor slim and powerful.

Two of the things which Italians do particularly well is building fine sports motorcycles and designing clothes. The Bimota DB1 was the most finely dressed motorcycle ever seen. It had a streamlined all-enclosed bodywork which gave the bike an image of total integrity. From head to tail-light, the curves flowed so smoothly there was no discernible point where the fairing met the tank and the tank met the seat. It was a perfect, rounded design concept. Never has a bike looked so fast at a standstill.

SUPER BIKE
Arai

BMW K1

The BMW K1 was a concept bike. The concept, like the bike, was German. It was a bike for traveling great distances, in comfort, at high speed: an *Autobahn* motorcycle. It was designed to replace the BMW R100RS as the ultimate sports tourer. Notwithstanding its bulk, the K1 was probably the most aerodynamic machine around, its bodywork designed to shield the prone ride completely from windblast and to minimize turbulence.

The K1, launched in 1989 for the 1990 season, was European. It had a feel that was far removed from any Japanese motorcycle despite being powered by a water-cooled in-line four-cylinder DOHC 987cc engine – a typically Japanese configuration. But the K1 motor, like all the K-series engines, was laid flat and, of course, the bike was shaft driven. It was also fuel-injected, a method of induction the Japanese seemed to shy away from at the time, but which the Europeans (Ducati, Bimota, Moto Guzzi) embraced.

It was the first BMW to have a 16-valve head – indeed, it was the nearest thing to a genuine sports bike which BMW had built, although sports tourer would perhaps be a better description. It felt different from a Japanese engine as it was tuned for a wide spread of power, redlining at a very conservative 8500rpm, where comparable Japanese motors had 3000rpm in hand. The power was restricted, deliberately, to 100hp. It vibrated too, a trait long since eradicated from Japanese engines. However, one innovation was its optional anti-lock brakes, a system which beat the Japanese to the market-place.

BMW made a great impression with their new top-of-the-range sportster, not least by painting the first models either bright red or blue and canary yellow. People saw you coming on a K1. Its weird removable seat hump caused comment, mostly unfavorable. Some considered the bike to be beautiful, many considered it ugly and cumbersome. Either way it was a bike which left a great impression.

Left and below: *The K1 was intended as a sports machine, but only in a limited sense. It was designed to be ridden fast, but in comfort – hence the giant fairing. It was perhaps the most effective streamlining ever to be used on a motorbike; arguably ugly, but effective and impressive all the same.*

Above: *The first bike off the production line at the new Triumph factory in Hinckley, England. The Trophy was a Japanese clone, but a sound base for the reconstruction of a British marque nevertheless.*

Above right and right: *After the motorcycling hype of the century, the Honda NR750, not surprisingly, failed to live up to the high expectations of a 'GP racer for the road.' The 125bhp road machine turned out to be soft and refined rather than wild and outrageous.*

Triumph Trophy 1200

The name 'Triumph' conveys more than a dozen books could do, so loaded with history, glory and tragedy is the old marque. This Triumph, though, is the cuckoo in the nest. The name has nothing to do with the old British make and is simply being used as a marketing gimmick, having been bought for hard cash from bankrupt stock.

Even so, the new British marque has much going for it and caused great surprise with its stunning first-born, the 1200 Trophy. Millionaire John Bloor, the man behind the project, built his bikes with no pretensions. The Japanese led the market, so he chose to copy them. He got it right first time. Never before had such a technically refined motorcycle been constructed in Britain, even if its specification was all too familiar.

The motor was a water-cooled four-cylinder parallel four laid across the frame. It had four valves per cylinder and double overhead cams. The factory claimed 140hp delivered at 9500rpm! Top speed was 153mph, attaining standing quarters in just 11.3 seconds. Its top speed was not quite as fast as the benchmark machine of the day, the 175mph Kawasaki ZZ-R1100, but in mid-range performance and acceleration, it was only marginally behind.

The bike must stand as a classic, having been built in secret from scratch, in a country which had given up on ever re-establishing a motorcycle industry of any note. It was launched in the face of the fiercest technological competition, muscling in to claim a valid, albeit small, share of the market. The only reservation for the Trophy was that it had all been done before, and slightly better, by the Japanese.

Honda NR750

Perhaps the most hyped bike of all time, the Honda NR750 finally made its debut in October 1991 after being displayed at various motorcycle shows for two years. Honda held back from releasing the bike for several reasons. It was the most technologically advanced bike ever designed and was therefore incredibly expensive to build. Should they sell it as such with a millionaire's price tag or make a cheaper version, more accessible to the public? Did they want the bike to find its way to the race tracks where it would undoubtedly dominate to the exclusion of all other marques? Was it projecting the kind of image which Honda felt reflected the direction of their company? Wasn't it just upping the performance stakes between the factories so much that they'd be accused of irresponsibility?

Honda agonized for two years while the public was tantalized to the point of frustration. Even if the public couldn't afford it, at least they'd like some news of the world's most exciting motorcycle to date.

What made it so special was the engine. The motor was a V-4, like the super-successful VFR750R sportsbike, but the pistons were oval, and each cylinder had eight valves. Effectively the bike was a V-8. To cope with the extra strain the connecting rods were made from super-strong and light (and expensive) titanium. Projected horsepower was 160 in tuned form, and the prototype was redlined at an amazing 15,000rpm. The chassis was a development of the VFR's aluminum beam frame with single-sided swingarm; the bodywork was plastic reinforced with carbon fiber.

It was the nearest a motorcycle could be to an engineering work of art and it had the price tag of an old master: £38,000 ($64,000)!

SUPER
BIKE
bimota

Bimota Tesi 1D

Bimota, the small innovative factory near Rimini, Italy, have never been frightened of trying something different. Their Tesi, launched in November 1990, was certainly that. It was the first production motorcycle with center-hub steering, a method of suspension and steering which uses a swing arm at the front as well as the back.

The idea is not a new one and Honda, in conjunction with Elf, raced their version at Grand Prix level during the 1980s. The man responsible for the Bimota effort was Pierluigi Marconi, a young and imaginative engineer who thought up the idea at college for his thesis – hence the bike's name; in Italian thesis is *tesi*.

The benefits of his chassis were manifold: it was incredibly light, comprising the two aluminum swingarms and two boomerang-shaped aluminum brackets which connected front to rear and held the engine. The engine, a Ducati 851, was a fully stressed chassis member.

The front suspension was damped by one shock absorber mounted horizontally alongside the front of the engine. Marconi claimed that with such a rigid chassis, he could achieve the correct degree of suspension damping with much less travel in the suspension unit. Therefore under braking, acceleration and cornering, the bike would be a great deal more stable than a conventional bike, having less front-end dive under braking and extension under acceleration.

The bike proved a difficult one to set up for the rider, as it was very sensitive to rider weight and riding style, but the factory rider could lap as fast on the Tesi as the YB4 racer it replaced.

The Tesi was famous for another reason too: in 1990 it was the most expensive production bike available, at a cool £25,000 ($42,000).

Left: *At the track, the author tests the Bimota chief engineer's university thesis – the Tesi 1D. The center-hub-steered bike proved supremely stable when braking from high speed and going into turns.*

Above: *The Bimota stripped bare. The Tesi would not be feasible without the extreme rigidity of the Ducati 851 engine which is a stressed member of the chassis.*

Below: *The Manx Norton, a thumping single with alloy barrel and head.*

Bottom: *One of the six original 'Manx' Nortons.*

'Manx' Norton Racing International

To the purist, the name Norton means just one type of engine: the four-stroke single. Its development spans 50 years from the cradle of motorcycling to the halcyon days of the British domination of the world market. Since J L Norton himself began an association of his marque with the Isle of Man TT in 1909, Nortons have been challenging for the Trophy, right up until the company ceased motorcycle production in the late 1970s.

The first 'Manx Norton' was actually called the Racing International and made its debut just as World War II broke out, at the Manx Grand Prix in September 1939.

Six special machines were prepared for selected racers. The new bike featured a 'plunger' frame, with twin shock absorbers, wrap-around oil-tank and megaphone exhaust. The engine was a single overhead cam with exposed valves and hairpin springs. The six were, in effect, prototypes for what would have been the 1940 model, had the war not intervened.

Of the six, five went back to the factory, but one was crashed and remained on the Isle of Man, forgotten in the panic of war. Now restored, it sits in a glass case in the Crosby Hotel and is the bike pictured here.

After the war racers were built based on the pre-war Manx contenders. They were known as the Manx Grand Prix racer, later simplified to Manx, and were built to 350cc and 500cc capacity (the model 40 and 30 respectively) with bore and stroke of 71 × 88mm and 79 × 100mm. Compression was 7.33:1. The head was aluminum cast to a bronze skull; the barrel was alloy with an iron liner and magnesium alloy crankcase. Gearbox was four-speed and close ratio.

It was a formidable machine and was developed over the next 15 years, providing stars such as Geoff Duke and John Surtees with machines to match their talents. In 1950 the 'featherbed' frame was introduced – fully duplex with swinging arm suspension - setting new standards of fine handling and stability. Manx Nortons would continue to win races up until its discontinuation in 1962.

Yamaha TZ750 Flat-Tracker

'They don't pay me enough to ride that thing.' Those were the words of World Champion Kenny Roberts as he walked away from the bike which gave him, in his own opinion, the greatest victory of his career. He was riding a TZ750-engine flat-track racer, an outrageous machine which was an unofficial Yamaha effort to redress the balance between Yamaha and the mighty Harley XR750s which had been giving Roberts and his vertical twin 750 such a hard time in 1975. He was in danger of losing his Number One plate.

The motor came from a factory prepared Yamaha TZ750 racer, a machine which produced 110hp at 10,000rpm. Top speed was in excess of 175mph. The motor's specification was a water-cooled two-stroke parallel four, bore and stroke 66.4 × 54mm giving 748cc. It was the forerunner of the OW31, in 1976, the fastest bike ever built.

Putting the evil motor into a flat-track chassis was the idea of works rider Steve Baker and his mechanic Bob Work. The idea began as a joke suggestion; after all the motor was producing nearly 50hp more than the four-stroke twin dirt racer, and nothing so powerful had even been conceived of as a flat-tracker. 110hp and no front brake! Besides, who would be able to ride the thing around a mile-long oval dirt track?

Kenny Roberts provided the answer. Over 25 laps of the Indianapolis Mile, the bike slid, snaked and weaved beneath Roberts as he desperately struggled to find traction. By the last lap he'd caught the roaring Harleys and blitzed by to clinch the most memorable victory of his career. The bike's success ended almost as soon as it had begun, cynically legislated out of the competition by the racing authority, the AMA. In doing so they ensured the TZ750 flat-tracker's entry into motorcycling legend.

Above: *The TZ750, a crazy idea for a crazy bike – an almost untamable engine in a spindly flat-tracker chassis had a brief reign of glory in the hands of King Kenny Roberts.*

Above: *One of the last works KR750s is pictured above. It is a bike prepared for G Bray for the 1968 season, but has no wins to its credit, just a cool $750,000 price tag.*

Right: *Suzuki RG500 taskforce, Brands Hatch, April 1977. World Champ Barry Sheene (7) waits for the starter flag with team-mates Pat Hennen and Steve Parrish in the Transatlantic Match race. The RG500 would carry Sheene to his second World Championship that year.*

Harley-Davidson KR750

Harley's road and dirt racer had perhaps the longest life-span of any racing bike. In the mêlée of competition, factories, racers and tuners constantly strive to improve their race bikes by redesigning, updating and incorporating any new idea which may just give an edge. Racing improves the breed, as they say. But it seems the Harley KR750 managed to race through its 18-year history without radical review of its original 1952 specification.

Was it so right first time? Hardly: the simple side-valve arrangement of the KR had been proved inferior by overhead valve machinery fifteen years before. It was actually producing less power than the WR racing machine it was supposed to replace! It differed in that the engine was of unit construction, whereas the W-series 45s had separate transmission cases. Its power may have disappointed initially, but the chassis was a great improvement, having a sprung rear end as an option. It also had hand-operated clutch, foot-operated gearchange and four-speed gearbox – all innovations – for Harley anyway. Less power it may have had, but it could be raced faster.

The real reason for its longevity on the American racing scene was, perhaps, the rather paternalistic attitude toward Harley-Davidson of the sport's organizing body, the AMA. When the fast and light British bikes from Norton, Triumph and BSA began running rings round the Harleys, the AMA's response was to ban all overhead valve engines. As side-valve motors were long-since obsolete, the ruling effectively left Harley to race against itself. Of course when Harley built the OHV XR750, the AMA obediently changed the rules to accommodate it.

Still, Harley developed the bike over the years and no-one could say the racing wasn't a stirring spectacle. Championships were as hard fought and glorious as any. Cal Rayborn, aboard a KR, was the first to lap Daytona at over 100mph in 1968. A KR ridden by Roger Reiman was clocked at an official time trial at 149mph the same year; amazing achievements on a bike whose design dated back to 1916. No surprise then that they became collectors' items – works KR750 models in good condition were selling for up to $750,000 in 1991!

Suzuki RG500

The name of Barry Sheene is inextricably linked with the Suzuki factory-production racer, the RG500. It was in its third year of existence that Sheene took the 500cc World Championship. The bike not only caused a stir because of its out-and-out speed – 185mph – but by its unusual design. It was innovation from spark plug to sump. In contrast to the across-the-frame parallel four YZR Yamaha, the layout of the water-cooled two-

Pat
PAT
TEXACO
HERON
BARRY
BATES
CHAMPION
SUZUKI
FORWARD TRUST
SUZUKI
FABERGÉ
SUZUKI
TEXACO
HERON
TEAM SUZUKI
BRUT 33
CHAMPION
TEXACO
HERON
TEAM SUZUKI
SUZUKI

Above: *The American Fred Merkel, riding a race-prepared Honda VFR750R, won the World Superbike Championship in 1989. The bike was handbuilt by Honda Race Corp, and came as a roadbike. Its design was based on the Honda RVF endurance racer.*

stroke was a square four cylinder, running two crankshafts. Bore and stroke were 56 × 50.5mm (made 'square' in 1976 at 54 × 54mm for a better midrange) and induction was through disk valve. Ignition was electronic.

Plumbing the exhausts proved a tricky problem and the result was two exiting forwards and back under the engine and two straight out of the rear two cylinders and under the tail unit. The carbs were mounted on the sides of the crankcases. Power for the 500 was said to be 90hp at 10,500rpm, with the powerband running from 9200rpm to 11,000rpm.

Credit for the design should certainly not all go to Makoto Hase of Suzuki, though, as the motor's design was basically four of the MZ 125cc single, built in East Germany by Walter Kaaden over 10 years before. The RG's success stems from the brilliance of his design.

The bike wasn't an instant success, suffering seizures, throwing Sheene from the saddle on occasion, and handling problems which caused weaving at over 150mph. In 1975 Sheene suffered a horrific crash when the rear wheel locked up while he was practising for Daytona aboard the Formula 750 triple. He was traveling at over 175mph. Despite broken bones throughout his body, his injuries including fractured vertebrae and kidney damage, Sheene was riding again in just seven weeks.

After a crash which would put anyone else out for the season at least, Barry Sheene won both the Dutch and Swedish Grands Prix just a few months into his 'recuperation.' By the 1976 season, the RG500's teething troubles were sorted out. After the previous year's unbelievable show of courage and determination, he amazed the world again, winning five out of the first six Grands Prix of 1976, securing the title. It was a feat he repeated in 1977 riding the RG500, giving him and Suzuki two consecutive world championships.

Honda VFR750R

The Honda VFR750R, otherwise known as the RC30, was built by Honda's racing arm, Honda Race Corporation, and was shown to the public

for the first time in late 1987. As a race bike it was a sensation, outclassing everything else on racetracks around the world at every level of the sport – with virtually no special race preparation. What made the RC30 still more remarkable was that its designers also intended the bike for road use. Not surprisingly, at the time it was lauded as the best sports motorcycle ever built.

The bike was based on Honda's factory four-stroke racer, the RVF750, a machine used to contend the world endurance championship. Thus the RC30 had a single-sided swingarm, patented by Elf and Honda, which meant the rear wheel could be removed by removing just one large nut, leaving the brake and drive chain in situ. It saved hard-won seconds during pit stops for the frequent tire changes necessary during 24-hour races. The front wheel too was quickly detachable.

The frame of the bike was all aluminum, a deep-section twin-spar design, the 90-degree V-4 engine crammed between the two main beams. The frame design came straight from the World Championship-winning GP racer, the NSR500.

The engine itself was based on the VF750 roadbike motor, but was fitted with a 360-degree crank, so the two forward pistons would rise as the two rear pistons fell. It gave the motor a raw edge and a distinctive droning exhaust note. To keep cam timing accurate at high revs, the cams were driven by gears, an expensive but worthwhile feature.

The engine was liquid-cooled and had DOHC. The bore and stroke were 70 × 48.6mm, giving 748cc. The barrels were cast as an integral part of the upper crankcase to aid stiffness and to make stripping the engine easier and quicker for the owner or mechanic. The compression ratio was an impressive 11.2:1.

A standard bike produced around 105hp at the back wheel, a race-kitted motor around 135hp. It would rev to 13,000rpm, but that extra performance did not come cheaply – around £10,000 on top of the original £8,500 purchase price. Not surprisingly such a high-specification bike was an instant success. With American Fred Merkel riding, the VFR750R won the 1989 World Superbike Championship.

Above: *The VFR750R's domination of the world's racetracks was undisputed, but it was also a phenomenal road-going sports bike, with qualities of roadholding and braking never seen before in a production machine.*

Above: *Jay Springsteen, Harley's No 1 rider of the 1970s, shows what an XR750 flat-track racer can do when the track isn't so flat.*

Opposite, top: *The definitive café racer, the 1958 DBD34. It was as threatening in the fierce competition of the racetrack as on the highway.*

Opposite, below: *A 350cc BSA Gold Star in the classic parade at the Isle of Man TT. In its day, the marque was a dominant force in racing on the island.*

Harley-Davidson XR750

In 1970, faced with an impressive display of Oriental technology, Harley-Davidson finally retired their KR sidevalve production racers after 15 years of racetrack success. The XR750 was its replacement. The bike was basically a sleeved down 883 Sportster – overhead valves (but larger), domed pistons, bore and stroke of 3.0005 × 3.291in and a four-speed gearbox. The cam profile gave a broad spread of power, 4800rpm – 6200rpm, producing 62hp.

It came as a dirt-track racer, a magnificent machine, stripped to bare essentials, its bulging motor dominating the spindly chassis. Its flat black and orange tank, short seat and tail unit and straight-pipe open exhausts made it look more like a weapon than a bike. Of course as a dirt racer it carried no braking facilities at the front, adding to its death or glory image.

Indeed Evel Kneivel, the flamboyant stunt man, drew some not-unwelcome publicity for the machine by adopting it as his mount for his acts of (often foolhardy) derring-do. Who knows how many he smashed up in front of ghoulish crowds.

The bike dominated the dirt-track ovals of America until only recently when suitable Japanese roadbike motors, like the Honda V-twin, were developed by less patriotic race teams, keen on squeezing Milwaukee from its number one slot. Names like Cal Rayborn and Jay Springsteen became motorcycling legends in their own time aboard XRs. In Rayborn's case, it was not on dirt, but on the high banking at Daytona with the XR motor in a road race chassis.

XR-based motors have powered Harley racers to glory as recently as 1986, in the Battle of the Twins. The motor was a tuned version of the XR1000 road bike, a model built to capitalize on the outstanding success of the XR750.

BSA Gold Star DBD34 Clubman

The BSA Gold Star was a racer. Whoever owned one would race it, whether it was for 30 laps of Brooklands or just to the next café on the highway. In pure competition the machine dominated for years, to the point where races looked like one-model events. The 1955 Junior TT saw 33 Gold Stars out of 37 starters roaring around the Isle of Man. BSA had that one sewn up for sure.

The Gold Star took its name from a racing achievement which proved great engineering skill and durability as well as rider courage. Before the war the British Motorcycle Racing Club awarded a gold-star lapel pin to any rider who managed an average race time of over 100mph at the Brooklands race track. On 30 June 1937 BSA fielded a specially prepared 500cc Empire Star running 13:1 compression and using alcohol fuel. The bike took Walter Handley to a win averaging 102mph, with a fastest lap of 107mph. He won his pin and the Gold Star was born.

War interrupted production and the Birmingham Small Arms (BSA) factory went back to its original business of making guns. After the war, the Gold Star began a reign which spanned trials, scrambles (motocross) and road-racing – not to mention unofficial café-to-café races.

The bike in its final incarnation, the DBD Clubman, was indeed a racer, built without compromise for that purpose. It was a single-cylinder machine with an almost-square bore and stroke of 85mm × 88mm. It featured a die-cast light alloy cylinder head and a barrel with separate rocker boxes. The deep finning on the barrel and head (to aid cooling) was like no other; it gave the motor a threateningly large presence and the rakish exhaust lent the machine a purposeful look which left no one in any doubt about its rider's intentions. Three-lap races of the Isle of Man and a 1½-inch Amal carb meant the tank had to be big. It was made of aluminum alloy, bare but polished, and held five gallons. A bike which could accelerate in top from 25mph to an easy 110 needed brakes too, and the Gold Star was offered with a 190mm full-width hub at the front.

It was an expensive bike to build and an expensive bike to buy and its glory days were numbered. There was a run of a 350cc version in 1959, but in 1962 the last Gold Star was built as BSA slipped into decline.

Marlboro
1
Marlboro
Marlboro

SHOEI
Marlboro
YAMAHA
YAMAHA
Marlboro

Above: *Steve Baker winning with the OW31 in the 1977 USA 750 championship.*

Above right: *The mighty engine of Baker's works OW31 for 1976 – a watercooled, two-stroke in-line four of tire-shredding power.*

Previous pages: *Wayne Rainey raced the YZR500 to two consecutive World Championships in 1990 and 1991, stamping his authority and that of Yamaha's race department on Grand Prix racing.*

Yamaha OW31

'It's the fastest bike ever made, there's no doubt about that. I'd say it would do 190 miles an hour. It's the best bike I've ever ridden.' That was factory rider Steve Baker's opinion of the Yamaha OW31 in 1976. It was an opinion shared by Suzuki's works rider and reigning World Champion Barry Sheene.

Five OW31s were built in 1976 and four of them made their debut at the Daytona 200-miler ridden by Kenny Roberts, Johnny Cecotto, Steve Baker and Hideo Kanaya. They qualified in the first four positions, with Roberts setting a new lap record in practice. In the race the bikes turned out to be too powerful for their own good, shredding the Goodyear tires, with only the young Cecotto hanging in there, perilously close to disaster. He took the victory, his rear cover worn down to the canvas. The next year the bike went into limited production – 70 bikes worldwide. In the Formula 750 class (which had just been given FIM world championship status that year) the bike was unbeatable.

The OW31 was a development of the TZ750, itself an enlarged capacity YZR500 Grand Prix racer. The water-cooled motor was a two-stroke parallel four, laid across the frame. It had seven-port induction, four 34mm carbs, six gears and a dry multiplate clutch. Bore and stroke were 66.4 × 54mm, giving 748cc capacity. Huge efforts had been made to save weight – magnesium was used in the engine and every nut and bolt on the bike was titanium. The bike was 20kg lighter than the production TZ750 at 136kg. Besides having terrific acceleration, the bike handled extremely well and used monoshock suspension, unlike the twin-shock TZ750.

The OW31 became an instant legend, among racers as well as race fans. Those privateers who managed to secure a bike in 1977 were virtually guaranteed success. Works rider Steve Baker clinched the Formula 750 championship that year, to no-one's great surprise after the OW31's stunning debut. It was, after all, undoubtedly the fastest motorbike in the world.

Above: *The beast unclothed. Rainey's 1990 bike represented the pinnacle of motorcycle development: 165bhp powering a 125kg bike.*

Left: *Kenny Roberts and Yamaha OW31, at the time, the world's greatest racer, riding the world's fastest bike, at the world's fastest track – Daytona, USA.*

Yamaha YZR500

The Yamaha YZR500 V-4 has several World Championships to its credit and, since it was developed by Eddie Lawson to winning form, has been proven to be the most reliable and all-round capable bike in the blue riband 500cc GP class. Lawson won three world championships on the YZR, – 1984, 1986 and 1988.

Only a handful of riders in the world were capable of riding the YZR to its limit. In 1990, when the Californian Wayne Rainey won his first Championship (he successfully defended it in 1991), the bike produced around 165hp and weighed just 125kg – about the same weight as a 125cc roadbike which has 25hp. Its top race speed was around 185mph, and it had a usable powerband running between 10,000rpm and 12,500rpm.

The engine was a two-stroke four-cylinder set in a V and placed in line with the frame. It used four 35.5mm Mikuni flat-slide carburetors and would do around 14 miles to the gallon! Fortunately it had a seven-gallon fuel tank. The gearbox was six speed, a cassette-loading type which could be changed in minutes to alter gearing. The exhausts were welded steel with carbon fiber end-cans.

The chassis was Yamaha's Deltabox aluminum frame with Swedish Öhlins suspension (widely accepted as the best available and adjustable to three million combinations!). Rainey could adjust the spring preload of the monoshock from the cockpit. In 1991 the bike was fitted with damping controlled by micro-processor, the so-called Öhlins Active suspension which changed the damping rates of the shock to suit conditions as the bike was being raced. The steering geometry was adjustable. The three-spoke wheels were made in Italy by Marchesini, cast in magnesium alloy. The brake calipers were made by the British firm Lockheed. Carbon fiber disks were used in the races. In the motorcycling world, it was simply the ultimate piece of engineering.

INDEX

Page numbers in *italics* refer to illustrations